Parenting and Child Safety Artificial Intelligence Resource Guide

Written By a Real Human

Tammy Toney-Butler

Dedication

First, God deserves all the glory and praise in my life. I am nothing without His grace and mercy. I was dead, a shell of a person, until He saw fit to set me free, transform my mind and body, and make me a new creature in Christ. This book is a Holy Spirit-inspired download, written as I embraced my new identity in Christ Jesus and fully stepped into the call God placed on my life as a Healing Evangelist. It is His masterpiece, and as you read it, please see Him as the author and not me.

Next, I want to thank David, my husband, who stayed with me and showed me what real love was about. A love that embraced the messy believed in my causes and steadied my troubled soul. A love that refused to quit on me when I was caught in an emotional flashback and full of the residue of trauma, dressed in shame-soaked, icky garments full of holes. David offered no judgment, just praise and unconditional love, providing hope that a girl like me, broken, could be worthy of love and sustain it. David empowered me to become the real me and supported me financially until I broke free of the past, and into all God purposed me to become as a warrior for His Kingdom.

Additionally, David's parents are Russ and Seda. Parents who loved me despite my messiness and showed me what it means to be part of a family. Always in my corner, full of unconditional praise and encouragement. Seda, always dressing me for every occasion, and most of all for giving life to their son, David.

To my mother, Dianne, I give thanks. A few weeks before she died, after we prayed, she permitted me to discuss our life, unfiltered, if it would save another family from being destroyed by generational trauma—a mother who made choices based on survival from a trauma-soaked lens—a mother whom I loved dearly despite her failure to mother me as I deserved. A mother who I know now is in heaven, with my baby sister Anita, and would be so proud of the woman I have become in Christ Jesus. My mother never stopped praising Jesus.

Now, to my father, Marcel, who never really survived Vietnam, and struggled with addiction, coping through alcoholism, and serving through his hidden pain, and helping many as a police officer. A father who lost his battle with complex PTSD suffered in silence until he died with a self-inflicted gunshot wound (suicide) when I was age fifteen, and we buried him on Father's Day. Please reach out for help to all those suffering in silence. The world is a much better place with you in it. A daughter always longs for her father, even if he cannot be one. I loved my dad despite his not being able to parent me as I deserved. I know he would have been proud of me. I was proud of him despite his messiness.

Next, to my baby sister, Anita, who went home to be with the Lord in 2025. She suffered much but loved much. She was my encourager, and despite our being apart for most of her life, separated by lost promises and broken environments, my love for her never ceased, as did my love for my other siblings. Trauma destroyed my family, and as I journey to keep it from destroying yours, I carry Anita with me. She was strong, despite her body failing her in the aftermath of trauma and coping through various addictions. She was homeless at times, trafficked as a child and adult, had an arrest record, and never spoke of all she endured. As they amputated lower limb by lower limb due to vascular issues, she was in and out of the hospital, and through it all, she remained devoted to and praising the Lord Jesus. I love you, Anita. I will see you again.

Furthermore, to all my brothers and sisters in Christ, I would be nothing without your prayers and love. United, we are stronger. We must be the Light!

Now, to every educator (schoolteacher, principal) who poured into me as a child and influenced my path. You know who you are, and I owe you so much that words could never convey!

Finally, to all the "thrivers" and "overcomers" who have made it and are sowing those seeds of hope. Keep it up, for the harvest is great, and the laborers are few. Keep shining your Light and stay strong as we navigate the darkness of this world. Find your voice and use it! In all things give thanks and pray without ceasing.

Introduction

Parenting in an AI-Generative Society can be a challenge, to say the least. Navigating chatbots, phones, and gaming consoles can be exhausting. What do you do as a parent to protect your children, shelter them, yet let them grow through safe use of technology? In the world of deep fake digital abuse, grooming through virtual platforms, and rising youth suicide rates, artificial intelligence (AI) consumption can be quite daunting. Do you resist its use altogether, or do you partner with AI for the good of society?

Moral and ethical questions where a parent must search their hearts, draw on past experiences, and seek the insight of others to determine a course of action, set boundaries by placing guardrails, and, as Christians, seek the Holy Spirit for guidance. This pocket guidebook is designed to help you educate yourselves on AI, work together with your children, and partner to ensure its safe use in your home. As we move forward, we will define AI, understand the risks, and know where to turn for help. This book is not exhaustive, not meant to be a comprehensive dive into AI use in children, but only a brush stroke, to start the conversation, and hopefully drive you to read the main book in this series, "Ensuring Humanity Is Not Lost in an AI Generative Society: A Christian's Guide to Artificial Intelligence."

As always, you must make your own decisions regarding AI use and your children, but do educate yourselves thoroughly on the topic, its dangers, and how to protect the most vulnerable from harm.

Chapter One
Definitions

Artificial Intelligence, as defined by the National Aeronautics and Space Administration (NASA), is an artificial system designed to think or act like a human, including cognitive architectures and neural networks; computer systems that can perform complex tasks normally done by human reasoning, decision making, creating, etc.[1] In addition, an artificial system designed to act rationally, including an intelligent software agent or embodied

---

[1] https://www.nasa.gov/what-is-artificial-intelligence/

robot that achieves goals using perception, planning, reasoning, learning, communicating, decision-making, and acting.[2]

IBM defines a chatbot as a computer program that simulates human conversation with an end user. Not all chatbots are equipped with artificial intelligence (AI), but modern chatbots increasingly use conversational AI techniques such as natural language processing (NLP) to understand user

---

2

https://www.federalregister.gov/documents/2020/12/08/2020-27065/promoting-the-use-of-trustworthy-artificial-intelligence-in-the-federal-government questions and automate responses to them.[3]

Examples of AI chatbots are ChatGPT, Character AI, Claude, Google Gemini, Meta AI, DeepSeek, Microsoft Co-Pilot, Grok, and Poe.[4]

3

https://www.ibm.com/think/topics/chatbots#
:~:text=A%20chatbot%20is%20a%20compu
ter,and%20automate%20responses%20to%2
0them.
[4] https://zapier.com/blog/best-ai-chatbot/

Chapter Two
Risks of Chatbot Use

Chatbot use among teens is becoming a daily habit for some, according to a survey by the Pew Research Center. In 2025, roughly 1 in 5 U.S. teens say they use TikTok and YouTube almost constantly, with 64% of the teens surveyed saying they use chatbots; 3 in 10 report daily use.[5] ChatGPT is the most widely used chatbot, with roughly 59%, according to the study, followed by Gemini coming at 23%.

---

[5]

https://www.pewresearch.org/internet/2025/12/09/teens-social-media-and-ai-chatbots-2025/

ChatGPT and other chatbot use do not come without risks. ChatGPT and chatbots, by design, want to draw you into a conversation and keep it going. Parental controls do exist and can help keep your teen safe if you remember to use them. Even though ChatGPT has improved its safety features and parental controls, significant gaps remain.[6]

In fact, two lawsuits by parents whose teenage boys died by suicide after engaging in an ongoing relationship with chatbots were discussed at a

---

[6] https://www.commonsensemedia.org/ai-ratings/chatgpt-5

Senate hearing.[7] A recent NPR episode discusses the dangers of ChatGPT use among adolescents, using the Bible App, as one example.[8]

Child psychologists also warn parents about using chatbots such as ChatGPT for homework assignments.[9] Highlighting the risks of losing critical

---

[7] https://www.c-span.org/program/senate-committee/parent-of-suicide-victim-testifies-on-ai-chatbot-harms/665660
[8] https://www.npr.org/2025/12/29/nx-s1-5646633/teens-ai-chatbot-sex-violence-mental-health
[9] https://www.childrens.com/health-wellness/ai-and-kids-what-parents-need-to-know

thinking skills.[10] Furthermore, a recent study found that ChatGPT encourages harm among teens.[11]

---

[10] https://med.stanford.edu/news/insights/2025/08/ai-chatbots-kids-teens-artificial-intelligence.html

[11] https://counterhate.com/wp-content/uploads/2025/08/Fake-Friend_CCDH_FINAL-public.pdf

AAP's New Approach to Safe Media Use

The American Academy of Pediatrics (AAP) developed a new approach to help parents and caregivers balance media and promote healthy habits.[12] The AAP recommends the 5 C's of media use.[13]

---

[12] https://www.aap.org/en/patient-care/media-and-children/center-of-excellence-on-social-media-and-youth-mental-health/5cs-of-media-use/

[13] https://www.healthychildren.org/English/family-life/Media/Pages/kids-and-screen-time-how-to-use-the-5-cs-of-media-guidance.aspx?_gl=1*1fpks9z*_ga*MTUyOTAzMjEzNi4xNzY5ODc4NDM1*_ga_F

5 C's of Media Use

Child
Content
Calm
Crowding Out
Communication

Recognizing that each child is different, and different ages require different techniques to ensure safe media use, the AAP developed different tips for

---

D9D3XZVQQ*czE3Njk4Nzg0MzQkbzEkZ
zEkdDE3Njk4Nzg0NjMkajMxJGwwJGgw

infants,[14] toddlers and preschoolers,[15] school-age children,[16] young teens,[17] and older teens.[18] Recommending each

---

[14] https://www.healthychildren.org/English/family-life/Media/Pages/infants-and-screen-time-5-cs-questions-to-ask.aspx

[15] https://www.healthychildren.org/English/family-life/Media/Pages/kids-and-screen-time-5-cs-questions-for-toddlers-and-preschoolers.aspx

[16] https://www.healthychildren.org/English/family-life/Media/Pages/kids-and-screen-time-5-cs-questions-for-school-age-children.aspx

[17] https://www.healthychildren.org/English/family-life/Media/Pages/kids-and-screens-the-5-cs-questions-for-young-teens.aspx

[18] https://www.healthychildren.org/English/fa

family develop a family media plan for digital safety.[19]

Chapter Three
Resources for Parents

https://www.commonsense.org/

https://www.healthychildren.org/English/healthy-living/emotional-wellness/Pages/finding-mental-health-care-for-your-child.aspx

---

mily-life/Media/Pages/kids-and-screens-the-5-cs-questions-for-older-teens.aspx

[19] https://www.healthychildren.org/English/family-life/Media/Pages/How-to-Make-a-Family-Media-Use-Plan.aspx

https://www.healthychildren.org/English/fmp/Pages/MediaPlan.aspx
https://www.healthychildren.org/English/family-life/Media/Pages/talking-with-teens-about-media-conversation-starters.aspx

https://counterhate.com/wp-content/uploads/2025/08/Fake-Friend_CCDH_FINAL-public.pdf

Chapter Four
Hotline Support

U.S. Support Lines

988 Lifeline

https://988lifeline.org/

Suicide and Crisis Helpline

To reach a helpline, you can call, text or chat with 988 Lifeline which is available 24 hours a day,

365 days a year. Conversations are free and confidential.

NEDA

https://www.nationaleatingdisorders.org/

Eating Disorder Support and advice

To reach a helpline call 800 931 2237 from 11am – 9pm ET Monday to Thursday, and from 11am

– 5pm ET on Friday. To access web chat support use this link between 9am – 9pm ET on

Monday to Thursday, and 9am – 5pm on Friday.

National Drug Helpline
https://drughelpline.org/

Substance Abuse Helpline

To reach a helpline, call the National Drug Helpline on 1-844-289-0879, which is available 24

hours a day, 365 days a year. Calls are free and 100% confidential.[20]

National Human Trafficking Hotline

1-888-373-7888

---

[20] https://counterhate.com/wp-content/uploads/2025/08/Fake-Friend_CCDH_FINAL-public.pdf

Report Human Trafficking in Florida, call 1-855-352-7233

1 (855) FLA-SAFE

Report Human Trafficking in Georgia, call 1-866-363-4842

National Center on Missing and Exploited Children Cyber Tip Line

1-800-THE-LOST

1-800-843-5678

National Sexual Assault Hotline

1-800-656-4673

National Teen Dating Abuse Hotline

1-866-331-9474

Runaway Hotline

1-800-786-2929

Domestic Violence Hotline

1-800-799-7233

Department of Homeland Security to Report Human Trafficking

1-866-347-2423

Need Training on Human Trafficking, Trauma, and Exploitation

Contact

Nurses United Against Human Trafficking, PA

www.nuaht.org

info@nuaht.org

The National Center for Missing and Exploited Children provides a free service known as *Take It Down*, which could help victims, who have possession of the image or video files, remove or stop the online sharing of nude, partially nude, or sexually explicit content that was taken while under 18 years old. For more information, visit https://takeitdown.ncmec.org

If you believe you are the victim of a crime using these types of tactics, retain all information regarding the incident (e.g., usernames, email addresses, websites or names of platforms used for communication, photos, videos, etc.) **and immediately report it to:**

FBI's Internet Crime Complaint Center at www.ic3.gov

FBI Field Office [www.fbi.gov/contact-us/field-offices or 1-800-CALL-FBI (225-5324)]

National Center for Missing and Exploited Children [1-800-THE LOST or www.cybertipline.org]

Reporting these crimes can help law enforcement identify malicious actors and prevent further victimization.[21]

21

https://www.ic3.gov/PSA/2023/psa230605

UK Support Lines

Samaritans

https://www.samaritans.org/

 Suicide and Crisis Helpline

To reach a helpline, call 116 123 for free, which is available 24 hours a day, 365 days a year.

Alternatively, you can send an email to jo@samaritans.org

and receive a response in several

days or you can organize a face-to-face chat with the organization here.

BEAT

https://www.beateatingdisorders.org.uk/

Eating Disorder Support and advice

To reach a helpline, use this link to find phone numbers for England, Scotland, Wales and

Northern Ireland, 365 days a year, 9am – midnight during the week and 4pm – midnight on

weekends. For 24-hour web chat support use this link.

Frank

https://talktofrank.com/

Substance Abuse Helpline

To reach a helpline for confidential advice, call Frank on 0300 123 660, available 24/7. You can

also text Frank a question to 82111 or send an email to frank@talktofrank.com. Alternatively,

Frank offers a live chat service from 2pm to 6pm, 7 days a week.[22]

---

[22] https://counterhate.com/wp-content/uploads/2025/08/Fake-Friend_CCDH_FINAL-public.pdf

About the Author

Reflective Spaces Ministry, Corp, is a 501(c)(3) non-profit founded in 2021 by Tammy Toney-Butler, a former emergency department nurse and sexual assault nurse examiner. Following the whisper of the Holy Spirit, she and her husband relocated to Lee County, Florida. They purchased a ten-acre parcel of land to begin a trauma-focused, healing ministry.

Tammy, a Healing Evangelist, can be found on the streets, going after the ones. Tammy's lived experience provides a unique teaching style and trauma-focused lens perspective, offering survivors environments conducive to healing mind, body, and spirit.

Tammy Toney-Butler, as a teenager, survived the loss of her father to suicide. She overcame being a victim of child sex trafficking and coping with the aftermath of trauma through various addictions through her faith in the Lord Jesus and is a powerful testimony of faith in action.

In 2023, Reflective Hour with Tammy Toney-Butler was launched in podcast and YouTube formats as a platform for transformational healing in Christ. The Reflective Spaces Ministry podcast was launched in podcast and YouTube formats in 2024.

Tammy is outspoken in her mission to provide a trauma-responsive pulpit and a compassionate, merciful lens through which one offers pastoral support. Love is her focus because the love of Jesus Christ heals all wounds, delivers, and transforms.

Tammy has become the mouthpiece for God's message of hope and healing worldwide. She is a published author whose works have been featured in the National Library of Medicine, Congress.gov, textbooks, and several professional journals.

Her memoir and healing devotional with journal pages are available on Amazon and Kindle.

Tammy has spoken at the United Nations, American Nurses Association (ANA) General Assembly, ANA New York, ANA Georgia, ANA Vermont, and has been a guest on network television.

From the ranch to the pulpit, from the trailer park to the assembly hall, God has moved mightily in Tammy's life, and the Lord Jesus Christ gets all the credit and honor for her transformation and restoration.

Reflective Spaces Ministry

Reflective Spaces Ministry, Corp, is a 501(c)(3) non-profit founded in 2021 by Tammy Toney-Butler, a former emergency department nurse and sexual assault nurse examiner. Following the Holy Spirit's whisper, she and her husband, David, relocated to Lee County, Florida. They purchased a ten-acre parcel of land to begin a trauma-focused, healing ministry.

The mission of Reflective Spaces Ministry is to provide reflective spaces for transformational healing and total restoration in a faith-filled environment for survivors of human trafficking, sexual violence, domestic violence, and childhood adversity to thrive. In 2025, our Founder, Tammy Toney-Butler, on her journey to wholeness, had an awakening as to the fundamental mission of Reflective Spaces Ministry.

Our goal is to show the heart of Christ to all we encounter, empowering and enabling them with the strength and courage required to look inward, reflect on the past adversities faced, and live transformed lives despite it. A reflective space within one's own heart, full of strength, power, and courage to face the dark, refuse to let it break them, and instead, process and overcome it one layer at a time. Bringing them to wholeness, physically, mentally, spiritually, and financially.

Consider donating to our direct survivor assistance programs, including day programs such as respite retreats for spiritual renewal through nature, as well as free healing ministry services.

The team believes in empowering every survivor with a safe retreat experience in a private setting while they process the dark and transition into the Light. Survivor empowerment is vital to recovery; assisting survivors/thrivers with transportation and living expenses as they transition on their healing journey is vital to long-term success.

Reflective Spaces Ministry assesses and meets them where they are on their healing journey, whatever that looks like for their unique situation. Finding gaps in existing services and bridging those ensures an overall positive transitional health experience. Overcoming barriers to care and supplying Maslow's Basic Hierarchy of

Needs ensures that youth transitioning into adulthood and vulnerable adults will not be left behind. "Throw-Away" youth are vulnerable to trafficking and exploitation; thus, wrap-around support is needed to bridge gaps and provide pathways for success.

Positive mentoring, role modeling, and empowerment are essential for this generation of young warriors to tap into their hidden purpose and become all God has called them to be in this stage of their lives. Positive childhood experiences (PACES), such as a day retreat at A&K Ranch, can help mitigate the effects of adverse childhood experiences (ACES). Empowering "thrivers" with resources and choices as they journey to wholeness is a mission worthy of your financial support.

Note from the Author

As a survivor of child sex trafficking and loss of father to suicide as a teenager, I know trauma and loss all too well. Our ministry, Reflective Spaces Ministry, focuses on hope and healing, one layer at a time. Through quiet reflection and inner work, one can achieve wholeness through faith in the Lord Jesus Christ, as I have done by His grace and mercy on my life.
As a Healing Evangelist, my heart is for the lost, broken, wounded, and those coping with trauma through various addictions. Love heals and

transforms, as evidenced by our Lord and Savior, Jesus Christ. My goal, His goal, is to empower, inform, and equip with the tools needed to thrive as maturity through faith increases, and deliverance is possible. I am not just interested in getting them out of Egypt but getting Egypt out of them. A renewed mind is possible, with ultimate healing in four key areas of fitness: spiritual, mental, physical, and financial. Achieving wholeness is possible if one puts the work in. Trauma destroyed my family, and I am on a mission to save yours and all those I encounter. Breaking generational patterns of abuse is an essential component of healing, trauma work, and a focus of our ministry. Thank you again for trusting me to deliver a healing message to you by way of His books.

Blessings and Peace,

Tammy Toney-Butler, Healing Evangelist

Sow A Seed: Donate

Consider SOWING A SEED to further our community outreach and evangelism efforts to spread the Gospel worldwide! Even if the ninety-nine are safe, we go after the one! Partner with our ministry by clicking on the ministry link below to help us gather the ones into the family of Light.

https://www.reflectivespacesministry.com/

https://www.paypal.com/fundraiser/charity/4406377

https://account.venmo.com/u/Reflectivespacesministry

Contact Information

Tammy Toney-Butler,

Reflective Spaces Ministry, Corp,
16295 S. Tamiami Trail, Suite # 133,
Fort Myers, FL 33908
info@reflectivespacesministry.com
www.reflectivespacesministry.com
www.reflectivespacespodcast.com

Reflective Hour with Tammy Toney-
Butler is available at:
www.reflectivehour.com

Tammy's Amazon Author Page to
Purchase her books:

https://www.amazon.com/stores/Tamm
y-Toney-
Butler/author/B0DC1VXP45?ref=ap_r
dr&shoppingPortalEnabled=true&ccs_i
d=f4bd4d22-5e2d-4bde-851e-
9127258b3bf8

Tammy is available to teach and empower women and men as they journey to wholeness through the Light and Love of Christ. Contact her to book an in-person prophetic healing session, meeting, service, or conference at www.tammytoneybutler.com

Nurses United Against

Human Trafficking

Nurses United Against Human Trafficking (NUAHT) was founded in 2020 by two nurses driven to abolish modern-day slavery. Dr. Francine Bono-Neri and Tammy Toney-Butler.

NUAHT offers education modules, membership resources, and consulting services for healthcare professionals, by building human trafficking protocols and community response teams.

The mission of NUAHT[23] is to eradicate human trafficking by raising awareness, providing education and resources, and participating in advocacy efforts, all for the hope of emboldening and empowering healthcare professionals, by establishing best practices and standards of care for this vulnerable and invisible population.

---

[23] www.nuaht.org

For every membership purchased, NUAHT donates to Reflective Spaces Ministry, a direct service provider for survivors of human trafficking, sexual and physical violence, and childhood adversity (trauma). Reflective Spaces Ministry provides all services for free.

Sinner's Prayer to be Saved

Dear Heavenly Father,

I come to you in the Name of Jesus. Your Word says, "The one who comes to Me I will by no means cast out" (John 6:37 NKJV). I know You won't cast me out or turn me away. I know You take me in and I am grateful for You and thank You. You said in Your Word, "Whoever calls on the name of the Lord shall be saved" (Romans 10:13 NKJV). I am calling on Your Name, now, Oh Lord, and I believe You have saved me, a lost sinner. You also state in Your Word, in Romans 10: 9-10, "If you confess with your mouth the Lord Jesus and believe in your heart that God has raised Him from the dead, you will be saved. For with the heart, one believes unto righteousness, and with the mouth of confession is made unto salvation." I believe Jesus rose from the dead for my justification. I am now reconciled to God. I confess Jesus as My Lord and Savior. Because Your Word says that "with the heart one believes unto righteousness," and I do believe with my heart, I have now become the righteousness of God in Christ (2 Corinthians 5:21). I now know I have been redeemed, restored, saved by the blood of Jesus.

Thank You, Lord Jesus. I praise and honor You and believe with this prayer and declaration that the Holy Spirit, Your Spirit, lives inside of me, making me fresh, clean, and new. I surrender to God's will for my life, instead of my will. Thank You, God, for giving me the heart and mind of Christ, for washing me clean, and setting me free.

If you prayed this prayer, welcome to the family! Please email us at info@reflectivespacesministry.com to discuss next steps and mail you out resources for your walk as a new Christian.

(Prayer adapted from Kenneth E. Hagin's Laying on of Hands Book, pg. 33, Rhema Bible Church.

Additional Books

(Tammy Authored or Was a Chapter Contributor)

• Toney-Butler, T. (2026). Ensuring Humanity Is Not Lost in An AI Generative Society: A Christian's Guide to Artificial Intelligence. Written by a Real Human. (1st Edition). ISBN: 979-8-9947264-4-0.

• Toney-Butler, T. (2026). Ensuring Humanity Is Not Lost in An AI Generative Society: A Christian's Guide to Artificial Intelligence. Written by a Real Human. (1st Edition). ISBN: 979-8-9947264-3-3.

• Toney-Butler, T. (2026). Ensuring Humanity Is Not Lost in An AI Generative Society: A Christian's Guide to Artificial Intelligence. Written by a Real Human. (1st Edition). ISBN: 979-8-9947264-2-6.

• Toney-Butler, T. (2026). When You Know That You Know That You Know, There Is a God: Memoir and Journal for Self-Reflection. (2nd Edition). ISBN: 979-8-9945907-4-4.

- Toney-Butler, T. (2026). When You Know That You Know That You Know, There Is a God: Memoir and Journal for Self-Reflection. (2nd Edition). ISBN: 979-8-9945907-3-7.
- Toney-Butler, T. (2026). When You Know That You Know That You Know, There Is a God: Memoir and Journal for Self-Reflection. (2nd Edition). ISBN: 979-8-9945907-2-0.
- Toney-Butler, T. (2026). A Journeying Journal to Wholeness Through the Lens of Faith. Devotional and Trigger Journal. (2nd Edition). ISBN: 979-8-9945907-9-9.
- Toney-Butler, T. (2026). A Journeying Journal to Wholeness Through the Lens of Faith. (2nd Edition). ISBN: 979-8-218-92010-4.

- Toney-Butler, T. (2026). A Journeying Journal to Wholeness Through the Lens of Faith. (2nd Edition). ISBN: 979-8-9945907-8-2.
- Toney-Butler, T. (2026). The Olive Press of Affliction: Crushing Season Field Guide. (1st Edition). ISBN: 979-8-9945907-5-1.
- Toney-Butler, T. (2026). The Olive Press of Affliction: Crushing Season Field Guide. (1st Edition). ISBN: 979-8-9945907-7-5.
- Toney-Butler, T. (2026). The Olive Press of Affliction: Crushing Season Field Guide. (1st Edition). ISBN: 979-8-9945907-6-8.
- Toney-Butler, T. (2026). Healing Is My Portion: Healing Scripture Field Guide. (1st Edition). ISBN: 979-8-9945537-1-8.

- Toney-Butler, T. (2026). Healing Is My Portion: Healing Scripture Field Guide. (1st Edition). ISBN: 979-8-9945537-0-1.

- Toney-Butler, T. (2026). Healing Is Your Portion Journal: Moving from Your to My; A Personal Journey Within. (1st Edition). ISBN: 979-8-9945907-1-3.

- Toney-Butler, T. (2026). Healing Is Your Portion Journal: Moving from Your to My; A Personal Journey Within. (1st Edition). ISBN: 979-8-9945907-0-6.

- Toney-Butler, T. (2026). Healing Is Your Portion Workbook and Journal: Moving from Your to My; A Personal Journey Within. (1st Edition). ISBN: 979-8-9945537-7-0.

- Toney-Butler, T. (2026). Healing Is Your Portion Workbook and Journal: Moving from Your to My; A Personal Journey Within. (1st Edition). ISBN: 979-8-9945537-2-5.
- Toney-Butler, T. (2026). Healing Is Your Portion: A Disciple's Guide to Divine Healing. (1st Edition). ISBN: 979-8-9947264-0-2.
- Toney-Butler, T. (2026). Healing Is Your Portion: A Disciple's Guide to Divine Healing. (1st Edition). ISBN: 979-8-9947264-1-9.
- Toney-Butler, T. (2026). Healing Is Your Portion: A Disciple's Guide to Divine Healing. (1st Edition). ISBN: 979-8-9945537-6-3.
- Toney-Butler, T. (2026). Healing Is Your Portion: A Disciple's Guide to

Divine Healing. (1st Edition). ISBN: 979-8-9945537-5-6.

• Toney-Butler, T. (2026). Healing Is Your Portion: A Disciple's Guide to Divine Healing. (1st Edition). ISBN: 979-8-9945537-4-9.

• Toney-Butler, T. (2024). A Journeying Journal to Wholeness Through the Lens of Faith. (1st Edition). ISBN: 9798300058586.

• Toney-Butler, T. (2024). When You Know, That You Know, That You Know, There is A God! (1st Edition). ISBN 9798334694101.

• Toney-Butler, T. (2024). When You Know, That You Know, That You Know, There is A God! Memoir and Workbook With Added Journal Pages for
Self-Reflection. M0D2077686515.

• Bono-Neri, F., Christopherson, T., Ernewein, C., & Toney-Butler, T. J. (2023). Healthcare Response. In J. W. Goltz, R. H. Potter, J. A. Cocchiarella, & M. T. Gibson (Eds.), Human trafficking: A system-wide public safety and community approach (2nd ed.).

• Earl, N., Fasold, L., Lares, T., Kadolf, T., Rose, C., Thomas, M., & Toney-Butler, T. J. Gibson (Eds.), Human trafficking: A system-wide public safety and community approach (2nd ed.).

• Toney-Butler, T. J., Lares, T., Edwards, J., Toal, P., Bakeman, B., Haba, L., Mayfield, T., O'Rourke, A., Kury, K., Klein, M., & Carey, R. (2023). Community and Faith-Based

Approaches. In J. W. Goltz, R. H. Potter, J. A. Cocchiarella, & M. T. Gibson (Eds.), Human trafficking: A system-wide public safety and community approach (2nd ed.).

www.ingramcontent.com/pod-product-compliance
Lightning Source LLC
Chambersburg PA
CBHW060354050426
42449CB00011B/2987